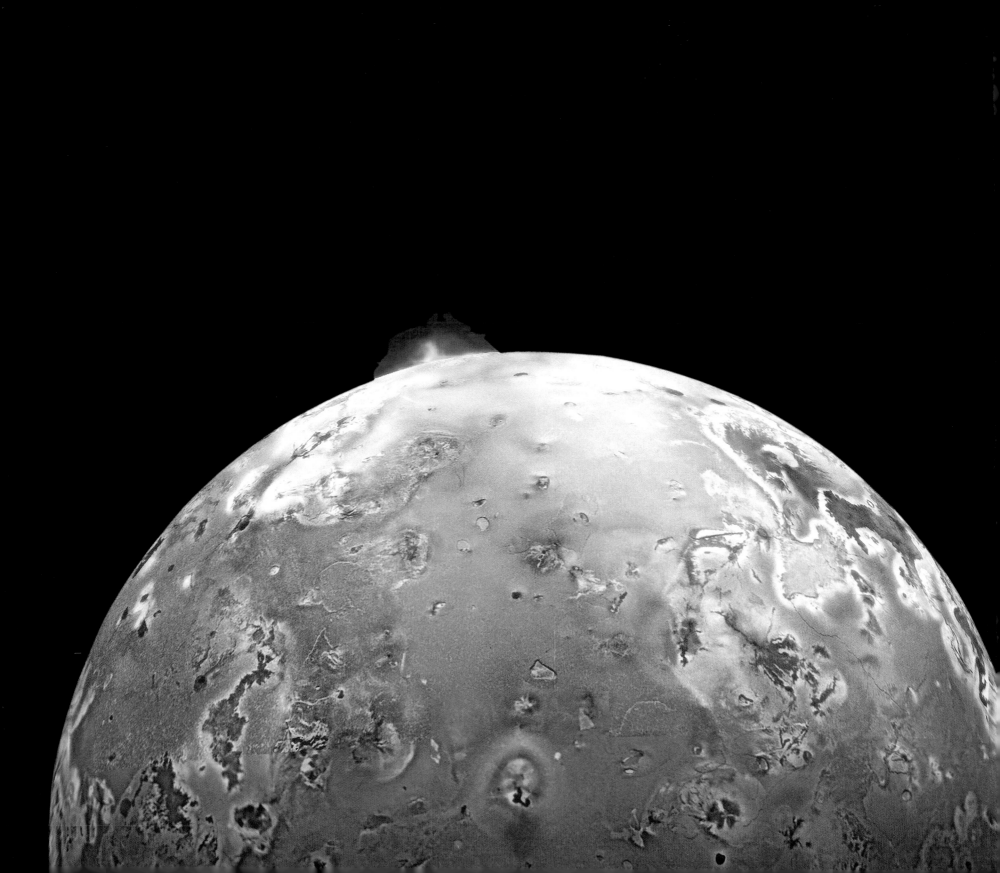

THE MOONS OF JUPITER

THE MOONS OF JUPITER

Kristin Leutwyler

Afterword by

John Casani
Chief Engineer for Flight Projects and Mission Success
and Original Project Manager for the Galileo Mission
NASA / Jet Propulsion Laboratory / California Institute
of Technology

A PETER N. NEVRAUMONT BOOK

W. W. Norton & Company
New York • London

For information about permission to reproduce selections from this
book, write to Permissions, W. W. Norton & Company, Inc., 500 Fifth
Avenue, New York, N.Y. 10110.

The text of this book is set in Meta
with the display set in Meta
Manufacturing by South China Printing Co., Ltd.

Jacket and book design by Cathleen Bennett

Library of Congress Cataloging-in-Publication Data

Leutwyler, Kristin.
    The moons of Jupiter / Kristin Leutwyler ; afterword by John Casani.
        p.   cm.
    "A Peter N. Nevraumont book."
    Includes index.
    ISBN 0-393-05060-2
    1. Jupiter (Planet)—Satellites. I. Title.
    QB404.L48 2003
    523.45—dc21   2003053974

W. W. Norton & Company, Inc., 500 Fifth Avenue, New York,
N.Y. 10110
www.wwnorton.com

W. W. Norton & Company Ltd., Castle House, 75/76 Wells Street,
London W1T 3QT

1 2 3 4 5 6 7 8 9 0

Created and produced by
Nevraumont Publishing Company
New York, New York
Ann J. Perrini, President

Contents

Jupiter and Its Moons

Bold Spirit! who art free to rove
Among the starry courts of Jove,
And oft in splendour dost appear
Embodied to poetic eyes,
While traversing this nether sphere,
Where Mortals call thee ENTERPRISE

—William Wordsworth, "To Enterprise"

On January 7, 1610, Galileo Galilei aimed a homemade telescope towards the heavens and chanced upon a spectacular new world—one that turned astronomy, and ultimately his life, upside down. On that clear winter night, though, he was just a tinkerer with a new toy. The telescope had been invented little more than a year earlier by a Dutch lens grinder named Hans Lipperhey. Galileo built his first copy—a vast improvement over the original—in July and by October had constructed an instrument that magnified distant objects 1,000 times over. "Forsaking terrestrial observations," he wrote in his famous pamphlet, *The Starry Messenger*, "I turned to celestial ones, and first I saw the moon .... After that I observed often with wondering delight both the planets and the fixed stars."

Jupiter, named for Jove, the Roman equivalent in myth to Zeus, immediately caught his eye. Galileo's primitive telescope didn't show the massive planet in anywhere near as much detail as the images on page 13, taken by the Cassini-Huygens spacecraft en route to Saturn in October, 2000. But it did well enough that Galileo, aided by his astonishing powers of observation, could distinguish at first three—and then four—objects around Jupiter that were somehow different from the host of background stars. "They aroused my curiosity somewhat by appearing to lie in an exact straight line parallel to the ecliptic," he wrote, "and by their being more splendid than others of their size." After a week of watching these "starlets," as he called them, Galileo realized he was looking at satellites that actually revolved around Jupiter. He named them the Medicean stars in honor of his patrons, the de Medici family.

The discovery of Jupiter's four large moons—now known as the Galilean satellites—refueled debates over competing models of the solar system: Aristotle's model placed Earth at the center whereas, in Copernicus's view, Earth and the other planets orbited Sun. Jupiter's moons didn't entirely disprove Aristotle's theory, but they certainly made the Copernican system seem all the more credible. "Here we have a fine and elegant argument for quieting the doubts of those who, while accepting with tranquil mind the revolutions of the planets about the Sun in the Copernican system, are mightily disturbed to have the moon alone revolve about the Earth and accompany it in an annual rotation about the Sun," Galileo wrote. Jupiter, too, he had discovered, traveled with companions on its orbital tour. It suddenly seemed far less peculiar that a single moon might circle Earth.

But the notion that Earth was not the center of the universe—and had planetary rivals with more plentiful moons—was heretical to many 17th century theologians. Galileo was dragged before the Inquisition, tried for his support of the Copernican model and sentenced to life imprisonment in 1633. He lived under house arrest until his death on January 8, 1642. More than 350 years passed before the Vatican cleared Galileo's name. On October 31, 1992, Pope John Paul made a formal declaration that fell just short of an apology. He conceded only that there had been a "tragic mutual incomprehension" between the astronomer and the church.

Little, in fact, had either the Vatican or Galileo comprehended just how serious a rival Jupiter and its moons are to Earth and Luna. They are more fairly compared to the entire solar system. We now know that Jupiter comprises two-thirds the mass of all nine planets and it is the biggest of the so-called gas giant planets, a class that also includes Saturn, Uranus and Neptune. Unlike the inner terrestrial planets—Mercury, Venus, Earth and Mars—these planets lack solid surfaces. Instead, they are made from the same stuff as stars: hot gas. This gas gives rise to the colorful storms that bloom in Jupiter's atmosphere on page 8. Inside stars, the gas is under such great pressure that it sparks a fusion reaction. Scientists estimate that, were Jupiter 50 to 100 times more massive, it would be a star like our Sun. Storm clouds on Jupiter travel in bands—jet streams that blow alternatively east and west—around Jupiter, whisking past each other like speeding lanes of traffic, which can be seen in the series of nine views of Jupiter on page 13. The planet's most famous storm, the Great Red Spot, makes an appearance in the third, fourth and last frames.

So, too, many of the Jovian satellites are more like planets than moons. The terrestrial planets have only three moons among them—Luna and Mars' satellites, Phobos and Deimos. But the gas giants are surrounded by swarms of smaller bodies. Jupiter has the largest entourage, with eight regular moons and more than 60 so-called irregular moons. These moons are categorically different from the terrestrial variety: Luna is probably a chip off Earth's block and the Martian moons may be captured asteroids. In contrast, the regular Jovian moons likely condensed from spare material around a proto-Jupiter, much in the same way as the planets coalesced from leftovers swirling near our newborn Sun.

For this reason, the Galilean moons contain many of the same ingredients found on Earth, varying in composition with distance from Jove. The closest Galilean, named Io, is primarily rock around a molten iron core. Hawaiian-style volcanoes explode from its blistered surface—one massive eruption appears on page 2—and its mountains rise higher than any found in the Himalayas. The next moon out, Europa, is also made from iron and rock, but with layers of water and ice on top. Cracks stained with what may be salts from a subterranean sea crisscross Europa's crust in the false-color image on page 4. If life exists anywhere beyond our planet in this solar system, a buried ocean on Europa is the most likely place.

Ganymede, the third Galilean satellite, has a thick ice mantle over rock and metal. Parts of Ganymede, once rocked by regular moonquakes, are faulted and wrinkled like Death Valley in California. Others are smooth, repaved by icy volcanism or additional tectonic activity. Some smooth areas may be akin to Earth's spreading centers, where new rock pushes up through the crust along the sea floor. Callisto, the outermost Galilean moon, seems to be a mix of rock and ice throughout. This moon's ancient face is heavily cratered and coated in a fine, dark dust. Callisto passes near Jupiter at the bottom of the picture on page 6 taken by the Cassini-Huygens spacecraft. Europa crosses below and to the left of Jupiter's most famous storm, the Great Red Spot. Callisto, like Europa, may possess a hidden sea.

For their resemblance to a mini-solar system, Jupiter and its moons afford a unique window into the birth and evolution of our world—from the origins of weather and geological change on Earth to the early chemistry of life itself. Countless astronomers have trained their instruments on Jove since Galileo's day and NASA has sent seven spacecraft past the planet to date, including the Hubble Space Telescope, the Cassini-Huygens spacecraft and the twins Pioneer 10 and 11 and Voyager 1 and 2. The most ambitious mission was named for Galileo himself. This spacecraft, which crashed into Jupiter in September, 2003, orbited the planet for eight years, collecting scores of data from 12 instruments.

The mission, like Galileo, the man, faced a number of serious challenges from the start. Its launch, initially scheduled for 1982, was delayed repeatedly by early problems with the space shuttle program and later, the *Challenger* disaster in 1989. When Galileo finally did launch on board *Atlantis* on October 18, 1989, it couldn't carry enough fuel to travel straight to Jupiter, some 934 million kilometers (580 million miles) away from Earth. Instead, the spacecraft was forced to take a modified path—one that sent it rather circuitously past Venus once and Earth twice to build momentum. Two years into the mission, shortly after the first Earth flyby, mission scientists discovered that the craft's main antenna had failed to deploy and later still, its recording device became stuck.

Despite these problems, the Galileo spacecraft chalked up a number of scientific firsts: it took the first close look at an asteroid named Gaspra; it discovered the first moon, Ida, around an asteroid, Dactyl; it made the first direct observation of a comet collision when Shoemaker-Levy 9 crashed into Jupiter in July, 1994; and it recorded the first global multi-spectral images of the far side of our moon. Near Jove, the Galileo spacecraft offered an unprecedented look at the mighty planet's starry court.

In the last few years of life, Galileo Galilei lost his sight entirely. "You may imagine the distress this causes me," he wrote to a friend in July 1637. "By my remarkable observations, the sky…the world…the universe…was opened a hundred or a thousand times wider than anything seen by the learned of all the past centuries. Now, that sky is diminished for me to a space no greater than that which is occupied by my own body." We can only imagine what Galileo might have thought had he seen photos from the spacecraft that bore his name.

Sounding the Clouds

In this photograph, spiraling clouds colored in shades of rose, olive and white, like veins running through smooth marble, blow across Jupiter's equator and fade into a cold blue horizon to the east, or right. The Galileo spacecraft took the shot at a range of about 1.2 million kilometers (745,645 miles) on November 5, 1996. These false hues are not only beautiful, they make it possible to visualize light scattered at three different wavelengths and thereby sound the depths of Jupiter's cloud decks—much as sonar uses reflected sound waves to plumb a column of water.

Galileo's Solid State Imaging (SSI) device recorded sunlight bouncing off Jove's atmosphere at 886, 732 and 757 nanometers. Like a sponge, methane in Jupiter's atmosphere absorbs different amounts of this near-infrared light, such that some wavelengths reach lower cloud layers than others. The methane soaks up most of the light at 886 nanometers, which therefore penetrates only the upper atmosphere, colored pink in this scene. The 732-nanometer light reaches intermediate cloud layers, shown here in a murky green. And the 757-nanometer light, unaffected by the methane, scatters off Jupiter's lower cloud decks, colored blue.

Galileo was the first spacecraft to look at the planet in near-infrared light. It was also first to directly sample the clouds. At launch in 1989, Galileo was comprised of two parts: an orbiter, a 2,223-kilogram (4,891-pound) vehicle scheduled to make 11 two-month trips around Jupiter and its moons, and a 339-kilogram (746-pound) probe, designed to plummet straight into Jupiter's atmosphere. The two components traveled together for 3.9 billion kilometers (2.4 billion miles) at an average speed of 70,811 kilometers (44,000 miles) per hour.

On July 13, 1995, the bullet-shape probe separated from the Orbiter. Five months later, it arrived at Jupiter, where it popped open heat shields for protection and a parachute for breaking. It began its descent at a speed of 170,000 kilometers (106,000 miles) per hour—the fastest impact speed of any man-made object to date. For 58 minutes, six instruments returned data by way of a radio signal to the Orbiter, which in turn relayed the information to Earth using the Deep Space Network, a collection of antennae that support all interplanetary spacecraft missions. (The antennae are strategically positioned at tracking stations in California, Spain and Australia to ensure round-the-clock, round-the-globe contact with spacecraft as Earth moves.) Shortly before the probe burned up, 200 kilometers (124 miles) down into Jove's stormy folds, it measured wind speeds as high as 724 kilometers (450 miles) per hour.

Jove's Tingling Nerves

"**Y**ou'll wait a long, long time for anything much / To happen in heaven beyond the floats of cloud / And the Northern Lights that run like tingling nerves." So wrote Robert Frost in the poem "On Looking Up by Chance at the Constellations" a safe 300-odd years after the Inquisition. In this false-color image, the Jovian equivalent to Earth's Northern display zings through the night sky, which is tinged green from light reflected by the planet's dayside glow. Jupiter's lit crescent, where this auroral show would fade in scattered Sunlight, falls out of the frame to the right.

The Galileo spacecraft photographed the scene on April 2, 1997 when it was approximately 1.7 million kilometers (1.05 million miles) from Jupiter. As on Earth, Jupiter's auroras appear where electrically charged particles from the solar wind jolt the atmosphere from above. These particles flow along magnetic field lines—loops of current that, like the wires in a whisk, wrap around the planet from pole to pole and trace the edges of its magnetic field, or magnetosphere. At the very highest latitudes, these lines extend almost straight up into the atmosphere and some, called open field lines, reach out into interplanetary space.

The wispy, white arc in this picture marks where these open field lines meet the closed field lines, which connect to the planet at either end. The charged particles at this border have traveled greater distances and therefore fly faster than those zipping down neighboring field lines. In keeping, they spark a brighter glow. The arc is part of a broad auroral ring, measuring hundreds of kilometers wide. Like a crooked halo, it dips to one side around the planet's pole of rotation and sags as low as 250 kilometers (155 miles) above the surface. The faint red aura that appears around the ring here is a signature of hydrogen in Jove's atmosphere.

In fact, Jupiter's atmosphere is 81 percent hydrogen, plus 18 percent helium and trace amounts of methane, ammonia, phosphorus, water vapor and hydrocarbons. This gas becomes denser with depth, like fog that becomes increasingly thick. At the planet's core, the pressure is immense. The hydrogen atoms are squeezed into a highly conductive, metallic liquid, unknown on Earth. Powerful currents build in this fluid as the planet turns, casting a magnetic net 10 times wider than the diameter of the Sun. If you could view Jupiter's magnetosphere from Earth, it would appear as large as our moon in the sky. It is by far the largest structure in our solar system.

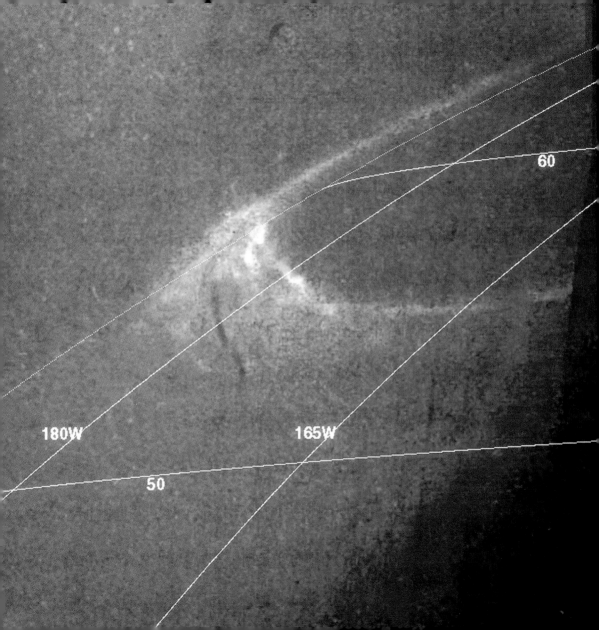

Blooming Storms

Jupiter's Great Red Spot blooms like a giant cabbage rose in this false color image, a composite of 18 photographs taken on June 26, 1996. The Galileo Orbiter snapped the pictures—six at a time through three different filters—within a span of six minutes. Combined they show thick folds of pink and white clouds in place of petals clustered around the eye of the storm. In the background, dark blue and green swirls recall the muted foliage in a Georgia O'Keeffe painting.

As in the photo on page 15, these colors correspond to different wavelengths of near-infrared light reflected by Jupiter's clouds. Because the wavelengths penetrate Jupiter's cloud decks to different depths, they give some measure of their relative heights. The Great Red Spot clearly churns in Jove's upper atmosphere: the pink areas are lofty thin hazes and white patches are high, thick clouds. The deepest clouds, colored black and blue like bruises, appear around the storm's edges and in the northwest, or upper left, corner.

The Great Red Spot is twice as wide as Earth, making it the largest storm in our solar system. It is also among the oldest. Astronomer Robert Hooke first spotted the hurricane-like high-pressure system in 1664 and, though it has changed shape, size and color since then, it shows no signs of waning. The busy storm completes one counter-clockwise rotation every six days and its winds reach speeds upwards of about 434 kilometers (270 miles) per hour. Scientists aren't sure what drives these screaming gusts. Steep temperature differences between the equator and the poles fuel terrestrial winds. But Jupiter's cloud tops are about -130 degrees Celsius (-200 degrees Fahrenheit) at all latitudes. One secret to the Great Red Spot's longevity may be that it never encounters land, which enervates Earth-borne systems.

In February 1998, the Hubble Space Telescope and the Galileo spacecraft documented an historic merger of two of Jupiter's long-lived storms. These so-called white ovals in the planet's southern hemisphere first appeared in the 1930s. The high-pressure storms gradually closed in on a low-pressure system between them until the three collided, creating a single, slightly larger storm. The marriage was far more violent than the hybrid system—the Perfect Storm of movie fame—that hit the northeastern United States in 1997. These white ovals were around 9,000 kilometers (5,600 miles) in diameter, or nearly three-fourths the diameter of Earth.

Seeing Red

Jove's temper was legendary—he threw lightning bolts to vent his rage—and in these two frames, taken 75 minutes apart by the Galileo Orbiter on October 5 and 6, 1997, hot white flashes erupt on the planet's night side near the equator. Thanks to moonlight from Jupiter's closest Galilean satellite, Io, it is possible to make out dim clouds in the background, colored here in shades of red. The spacecraft's Solid State Imaging (SSI) system took the photos at a range of 6.6 million kilometers (4.1 million miles).

The picture on the left shows lightning storms at two different latitudes. The image on the right reveals three separate storms. To capture these flashes, mission scientists left the SSI's shutter open for about one minute, recording multiple strikes at a time. Jovian lightning has a similar flash rate to Earth's bolts, but the flares are brighter and larger. The storms measure one or two thousand kilometers (621 to 1,243 miles) across and the flashes span hundreds of kilometers. These bursts probably occur at depth within Jupiter's cloud layers, where there is more water. They light up ammonia ice high in the atmosphere.

Like Jove, Galileo had a temper and his troubles with the church were not all that had him seeing red. Sparks flew in 1614 when Simon Marius—or Mayr in his native German—claimed to have discovered Jupiter's moons before Galileo. Marius, once a student among Galileo's followers at the University of Padua, made the assertion in a book called *The Jovian World, discovered in 1609 by means of the Dutch Telescope*. He maintained that he had first spotted the four moons in November 1609 and completed his studies on December 29 that same year. Marius used the Julian calendar and this date in the Gregorian calendar was January 8, 1610—two days before Galileo stated he began his observations.

There is no way to know whether Marius independently discovered Jupiter's four big moons during the very same week as Galileo did, but most scholars agree that it seems unlikely. For one thing, Marius did not describe in any detail, as Galileo had, the actual movements of these satellites. And if Marius did lie, it wouldn't have been the first time. In 1607, one of his students published under his own name an instruction manual on the sector originally written by Galileo.

Galileo vented his anger towards Marius in 1623: "This same fellow [Marius]...did not blush to make himself the author of the things I had discovered and printed in that work [*The Starry Messenger*]," he wrote, but "you may see how he himself, through his carelessness and lack of understanding, gives me in that very work of his the means of convicting him." Marius never did pay any price for his plagiarism, having resettled in Germany beyond the reach of Italian law.

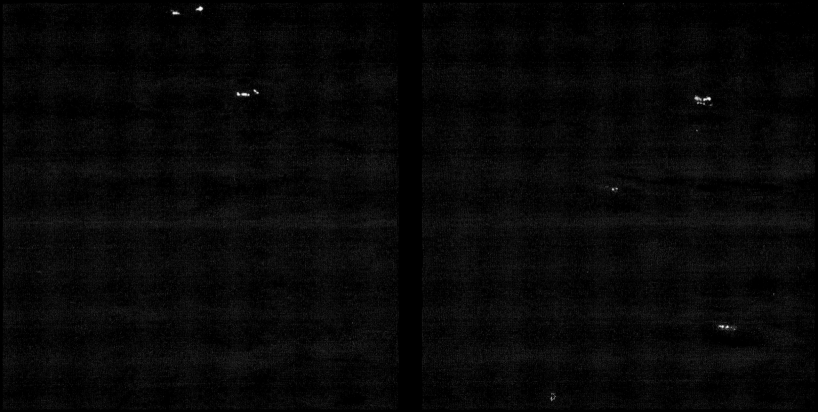

Starry Night

Were this photograph not taken 90 years after Vincent Van Gogh painted *The Starry Night*, you could almost imagine that it offered him inspiration. Jupiter's Great Red Spot, an enormous brown seed in the upper right corner, lies exactly where Van Gogh placed the biggest, brightest star in his composition. Rippling clouds, like curling smoke, pass below the Great Red Spot from the lower left to the upper right, tracing the path of slate blue mountains in the painting's background. And a wake of white turbulence left of the Great Red Spot echoes the bright brushstrokes that swirl across Van Gogh's masterpiece from the upper left to the middle of the sky.

In fact, the Voyager 1 spacecraft snapped this scene on March 1, 1979 through a variety of color filters, combined here to dramatic effect. The spacecraft was then approximately 5 million kilometers (3 million miles) from Jupiter and so the smallest visible details are about 55 miles (95 kilometers) wide. Voyager 1 ultimately passed within 206,700 kilometers (128,400 miles) of the planet that March. Its twin, Voyager 2, flew within 570,000 kilometers (350,000 miles) of Jupiter in July.

These two spacecraft, both launched in 1977, took advantage of an uncommon arrangement of the solar system's outer planets—one that occurs only once every 175 years. Beginning in the late 1970s, Jupiter, Saturn, Uranus and Neptune lined up such that the Voyager craft could efficiently zip from one to the next. Voyager 1 flew past Jupiter and then, using the planet's gravity like slingshot to gain acceleration, sped on to Saturn. From there, it veered off the ecliptic—the plane in which most of the planets circle the Sun—and headed for interstellar space. Some nine months after Voyager 1's visit, Voyager 2 flew past Jupiter and then Saturn—and then Uranus and Neptune, swinging from one gas giant to the next like Tarzan on a jungle vine.

Voyager 1 and 2 are still in operation. Both have seven functioning instruments, five of which are returning data via the Deep Space Network for dedicated scientific teams. As of February 17, 1998, Voyager 1 overtook the Pioneer 10 spacecraft and became the farthest human-made object in space. As of March 2002, it was about 12.6 billion kilometers (7.8 billion miles) from the Sun. Voyager 2 was not far behind, at 9.8 billion kilometers (6 billion miles) from the Sun. Should these vessels ever encounter life beyond our solar system, each carries a 12-inch gold-plated copper disc on which scientists recorded a variety of terrestrial sights and sounds. The late Carl Sagan chaired the committee that selected the disc's contents, including 115 pictures, music and spoken greetings in 55 languages.

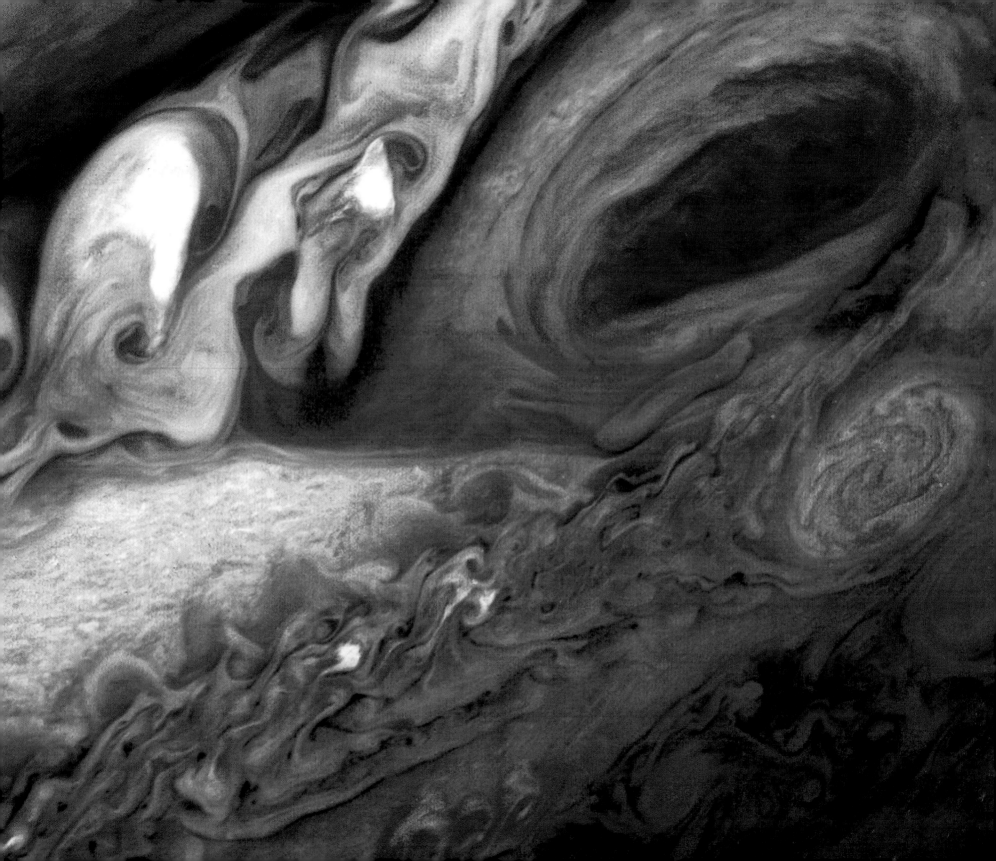

Between the Clouds

When Galileo's kamikaze probe plunged into Jupiter's atmosphere in July, 1995, it didn't find what exactly scientists were expecting. Models had predicted that the probe would encounter thick layers of clouds made from ammonia ice, ammonium hydrosulfide crystals and water, in that order. But instead it passed through only thin hazes and detected no water at all. Measurements from the probe's six instruments further indicated that Jove's innards were much hotter, dryer and windier than anticipated. Researchers quickly guessed that the probe had entered the atmosphere through one of Jupiter's so-called hot spots, clearings in the planet's cloud cover where radiant energy from the interior shines through.

In many respects, it was a lucky break. These hot spots, which are about the size of North America and last for several months at a time, are not common. They typically alternate with plumes of thick clouds in the planet's equatorial belt and, all tolled, account for less than one percent of Jupiter's area. Had they tried, scientists couldn't have directed the probe at such a tiny mark. But even so, it meant that the probe had missed its chance to sample more ordinary Jovian weather.

One such hot spot, colored dark blue, appears in this visualization of the cloud layers at Jupiter's equator, constructed using near-infrared images taken by the Galileo Orbiter on December 17, 1996 from 1.5 million kilometers (about 930,000 miles) away. The picture looks west towards the hot spot from within Jupiter's clouds, simplified into two discrete layers. The lower layer is false-colored and assigns red, green and blue shades to 756-, 727- and 889-nanometer reflections. The light blue clouds are high and thin; the white clouds are high and thick; and the reddish clouds are the lowest in this scheme.

The Galileo probe's measurements helped scientists gain a better understanding of Jupiter's odd hot spots. They now think that these holes in the cloud layer are sites of sudden and dramatic downdrafts. The air currents moving west to east just north of Jupiter's equator also ride a steep roller-coaster course up and down—from levels of low pressure to high pressure to low again. Researchers believe that water and ammonia in the atmosphere condenses into clouds that rise in the form of downy equatorial plumes; these clouds are then sucked down over 100 kilometers (60 miles) or so and wrung dry, creating clearings; once past the hot spots, though, the air rises again, forming more thick clouds.

Magnetic Footprints

Like high-speed jets hitting a hard surface, streams of electrons, protons and ions crash into Jupiter's atmosphere in this scene, creating auroral splashes of fluorescent blue light. Upon impact, the particles excite gas atoms, making them glow like the gas zapped in neon signs. Because Jupiter's atmosphere is primarily made from hydrogen, its auroras, present at both poles, would appear red to the naked eye. To make the lights easier to see, though, the Hubble Space Telescope viewed them through ultraviolet filters on November 26, 1998.

On Earth, energetic particles from the Sun fuel polar aurorae. But Jupiter's three largest moons—Io, Ganymede and Europa—also blast it with particle streams, sparking telltale emissions called magnetic footprints. The footprints appear here as bright focused lights within the halo over Jupiter's North Pole: Io's footprint falls off to the left; Ganymede's lies near the middle; and Europa's rests just below that and to the right. These streams flow within so-called flux tubes—vast channels of current that connect each of the moons electrically to Jupiter's poles. Researchers linked the magnetic footprints to the flux tubes when they noticed that the spots remained relatively fixed as Jove spun. The rest of the auroral halo pirouettes with Jupiter at its breakneck pace: one day on the gas giant lasts for 10 short hours.

Io's footprint, which measures about 1,000 to 2,000 kilometers (600 to 1,200 miles) across, is particularly vibrant. In this photo, it almost looks like a comet blazing overhead. In fact, were you hovering near Jupiter's cloud tops, Io's footprint would flood the sky in an instant, passing from east to west at a speed of about five kilometers per second (10,000 miles per hour).

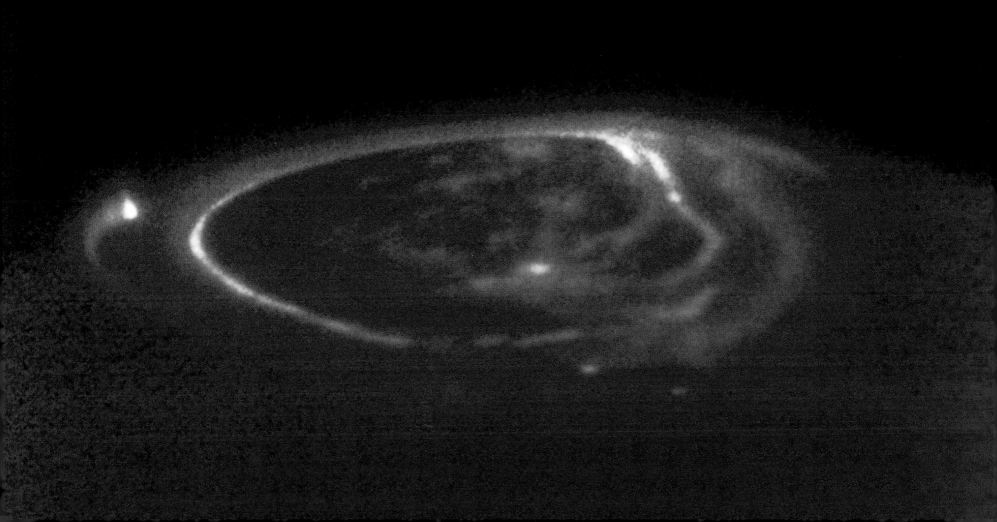

Jupiter's Paramours

Though Simon Marius' claims at discovery were questionable at best, the names he proposed for Jupiter's four big moons eventually stuck. They, too, were not an original idea, but at least Marius admitted it: "Jupiter is much blamed by the poets on account of his irregular loves," he wrote in *The Jovian World*. "Three maidens are especially mentioned as having been clandestinely courted by Jupiter with success. Io, daughter of the River, Inachus, Callisto of Lycaon, Europa of Agenor. Then there was Ganymede, the handsome son of King Tros.... This fancy, and the particular names given, were suggested to me by [Johannes] Kepler, Imperial Astronomer, when we met at Ratisbon fair in October 1613."

The names did not immediately catch on. Galileo had simply referred to the moons by number, starting with one for Io and so forth, and scientists did the same when they discovered the first five satellites of Saturn, beginning in 1655. But a problem arose in 1789 when William Herschel, who discovered Uranus eight years earlier, found two new moons closer to Saturn than the moon that had, up until then, been known as number one. To avoid renumbering by distance or discovery date, Herschel's son offered a simple suggestion: name the moons for mythological figures. When another moon appeared in 1848, its discoverers picked up on the idea, christening Saturn's satellites for the Titan's brothers and sisters. Soon thereafter, astronomers adopted Marius and Kepler's monikers for Jupiter's moons.

In this photo, assembled from three black-and-white images taken by Voyager 1 on February 5, 1979, three of the large moons—collectively named for Galileo and individually named by Marius—can be seen. Io is colored yellow and brown and appears slightly above and to the right of the Great Red Spot. Shiny Europa, its face covered with highly reflective ice, stands off to the far right. Callisto is barely noticeable against the dark sky at the bottom of the frame, though it is nearly twice as bright as Earth's Luna. All three moons orbit Jupiter in the same, equatorial plane, but they appear staggered in this photo because Voyager 1 viewed them from above.

It's a good thing astronomers adopted mythological names for Jupiter's moons before modern times. Between 1892 and 1974, scientists found another nine moons around Jupiter. Voyager 1 and 2 spotted another three in 1979. From 1999 to March 2003, some 30 Jovian moons made their debut (see page 217)—and there may be more soon coming. Jupiter certainly had numerous liaisons but Marius' convention of naming the moons for these trysts is strained. Some of the newer moons are named simply for other women in Jupiter's life—including a she-goat that suckled him as a babe (see page 46).

Io in Transit

The Starry Messenger contains several sketches by Galileo's hand preserving what he saw through his telescope in the beginning of 1610. Some show curved patches of shadow and light that the astronomer correctly took for rugged terrain on our moon. Others note the arrangement of stars in various constellations, including Orion's Belt and Sword and the Pleiades. And yet a third set of pictures—lines of two, three and four asterisks around a larger letter 'O'—document the movement of Jupiter's moons as Galileo watched them day after day.

On January 7th, he saw two "stars" to the east of Jupiter and one to the west; the next day all three were to the west; two days later, only two moons appeared to the east. Three days after that, he saw all four satellites—one to the east of Jupiter and the others to the west. He charted the moons' progress in this manner—measuring their distances from Jupiter, each other and stellar landmarks—until March 2nd, some 10 days before his book was published. Among his observations, Galileo wrote that "the revolutions are swifter in those planets [moons] which describe smaller circles around Jupiter, since the stars [moons] closest to Jupiter are usually seen to the east when on the previous day they appeared to the west, and vice versa."

In this series of images photographed by the Hubble Space Telescope on July 22, 1997, the Galilean moon closest to Jupiter whizzes from one side of the planet to the other. Galileo was right that Io moves fast: less than two hours elapsed between the first frame on the left and the last. Compared to Luna's 28-day orbit around Earth, Io completes one full rotation in only 1.8 days. Because the Hubble Space Telescope took these shots through ultraviolet and violet filters on the Wide Field and Planetary Camera 2, the colors are nowhere near what we would normally see. Io looks brown and white, glistening with sulfur frosts, against a cool, silver-gray Jupiter. The image is sharp enough to make out features as small as the state of Connecticut, or about 150 kilometers (93 miles) across.

As Io passes between Jove and the Sun, it casts a perfectly round, black shadow on Jupiter's clouds visible in the second and third images. Because Jupiter and Io are so far from the Sun, this shadow, which appears just to the right of Io, is basically life-size: it measures some 3,640 kilometers (about 2,262 miles) across, roughly the same diameter as Io. By comparison, the path of totality during a solar eclipse on Earth is never wider than 269 kilometers (167 miles). The shadow zips across Jupiter at a speed of 17 kilometers per second (38,000 miles per hour).

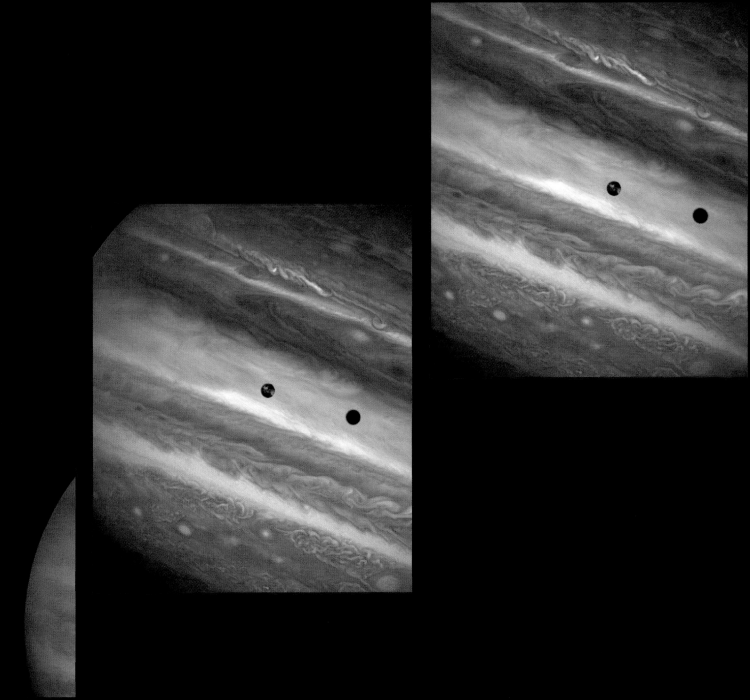

A Parade of Plumes

In one of Aesop's fables, Jupiter summoned all the species of birds so that he might choose a leader among them. The homely jackdaw—a runt cousin of the rook and crow—knew that, based on appearance, he wouldn't win. So the notorious thief ventured into the woods to collect more colorful plumes fallen from his feathered brethren's wings. When the time came to appear before Jove, the jackdaw dressed in his most fabulous finds—feathers of every shade, shape and size. Naturally, Jupiter proposed that the unusual bird rule the rest, but they quickly protested: the angry flocks reclaimed their feathers and left the jackdaw bare. Still, Jupiter did not change his mind. As Ambrose Bierce tells it in his book *Fantastic Fables*, "'Hold!' said Jupiter, 'This self-made bird has more sense than any of you. He is your king.'"

In this photograph, colorful Io, decorated with volcanic plumes, parades in front of Jupiter's southern hemisphere. The moon is, like the jackdaw, largely self-made in its looks: the hints of yellow, orange and red seen here are thought to be sulfur compounds deposited on the satellite's surface by way of regular eruptions. Voyager 2 snapped this photograph on June 25, 1979 from about 12 million kilometers (8 million miles) away. At this range, the smallest visible features on Io and Jupiter are about 200 kilometers (125 miles) wide—too big to see individual volcanic eruptions. Only a few days later, though, beginning on July 5, Voyager 2 began photographing Io's spectacular plumes. It continued tracking eruptions for several days after its closest pass of Io on July 9th.

In many ways, Io was the ruler of the Galilean moons during the Voyager trips. Perhaps the single most important discovery of the mission was that active volcanoes erupted on Io's surface. In all, Voyager 2 found eight smoldering plumes—one fewer than Voyager 1 had discovered during its Jovian pass four months earlier. By the time Galileo finished its studies of the moon, scientists had identified a total of 120 Ionian hot spots.

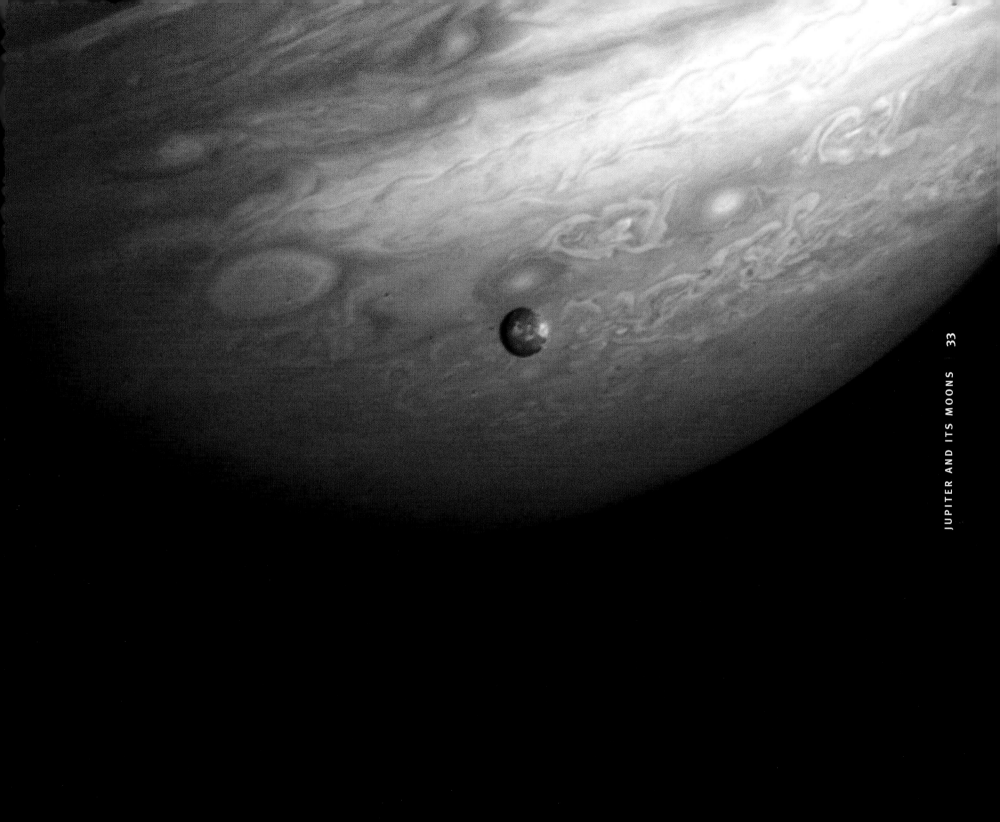

Europa From Afar

If volcanic Io ruled the imagination during Voyager's flybys, then Europa—suspected of hiding a subterranean sea—reigned supreme during Galileo's tour of duty 20 years later. After the spacecraft's primary mission ended in December 1997, NASA received funding for an additional two years and 14 orbits of study, called the Galileo Europa Mission (GEM). During GEM, Galileo came close enough to Europa to see features as small as a school bus—among them rafting ice floes and salt-stained hills (see images beginning on page 111). But it was not close enough to spot any signs of life. Not that life would be easy to see: before Galileo arrived at Jove, mission scientists looked for life on Earth as an exercise. Apart from oxygen and methane in the atmosphere—and strong infrared signals from land masses—unusual radio emissions offered them the only clue that Earth was home to intelligent beings.

Here mysterious Europa lurks off to the right, casting a black shadow on Jupiter's face below and to the left of the planet's center. Like Io's shadow, this one is nearly the same size as the moon at about 3,100 kilometers (1,926 miles) in diameter. The Cassini-Huygens spacecraft snapped the images used to construct this photograph from a distance of 81.3 million kilometers (50.5 million miles) during its Jovian flyby in 2000, shortly after GEM ended. It took the shots in blue, green and red light so that the combination very nearly approximates what our eyes would see.

Scientists are eager to see what Cassini discovers when it arrives at Saturn in June 2004. Like the Galileo mission, the Cassini-Huygens mission is named for the men who discovered the first of Saturn's 30 known satellites: Christiaan Huygens found the first and largest moon, Titan, in 1655; Giandomenico Cassini noted another four, two at a time, in 1671 and 1684. And like the Galileo craft, this one, too, is comprised of an orbiter—to investigate Saturn's atmosphere, rings and magnetosphere—and a probe that will hopefully land on Titan, an icy moon that shares several traits with Jupiter's satellites.

Voyager 1 passed within 4,000 kilometers (2,486 miles) of Titan, which was then considered the largest moon in the solar system. In fact, its width was misjudged: Titan is cloaked in a thick, opaque atmosphere that pads its girth. In reality, it is slightly smaller than Ganymede. Titan is the only satellite in our solar system that has such a thick atmosphere, made mostly from nitrogen and trace amounts of argon, methane, water and organic compounds. This smog may be quite similar to the early chemistry on Earth. Observations made by the Hubble Space Telescope have revealed what appears to be an enormous, bright continent on the Titan's surface, which may also host an ocean of ethane.

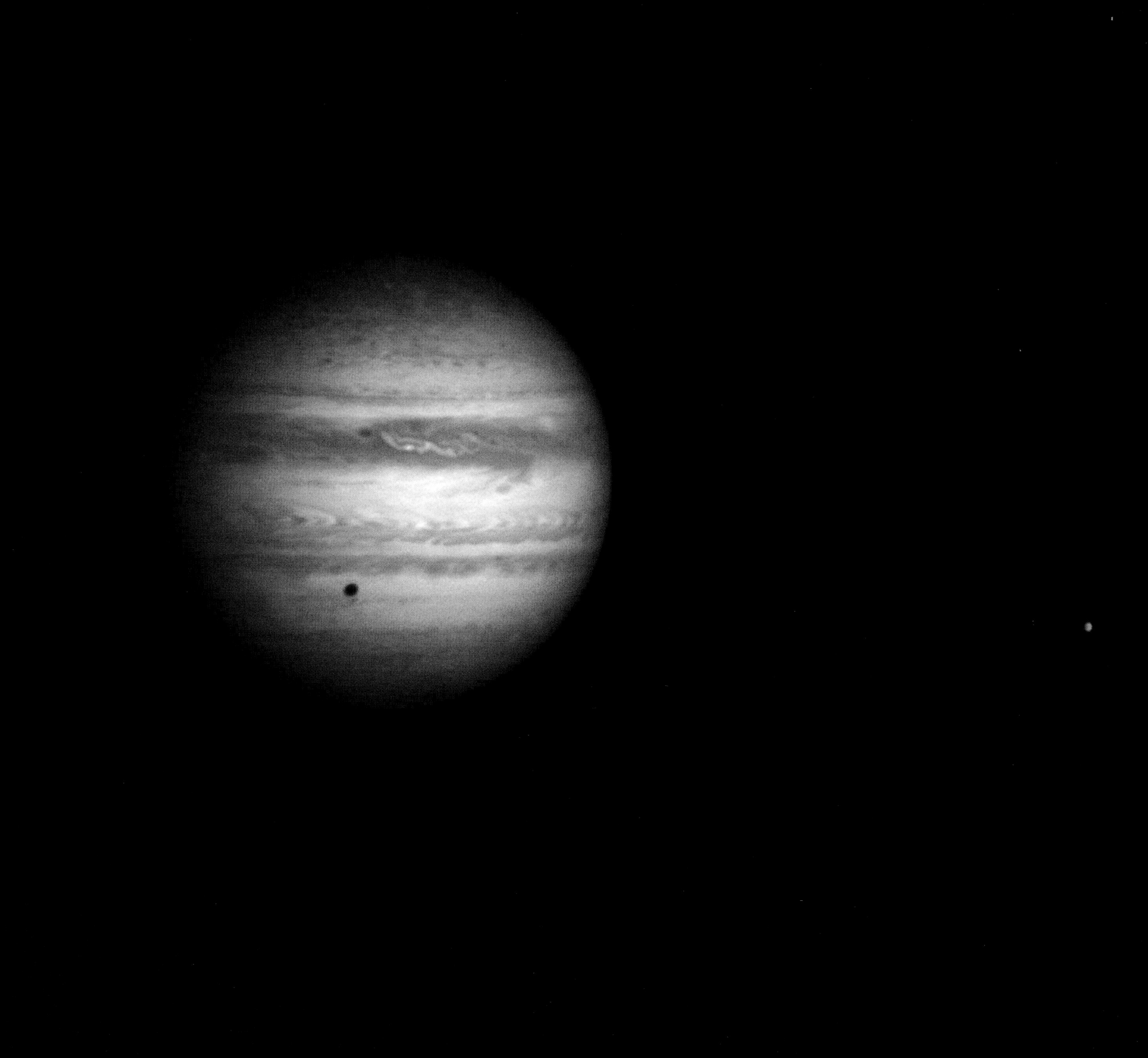

On a Hot Spot

Like the photo on the previous page, this one shows Europa's shadow sailing across Jupiter's face. An enormous black coin, where the moon blocks out the Sun's light, hovers in the lower left corner. But unlike the other image, this photo is false-colored to highlight temperature and height differences in Jupiter's various cloud layers. Red reveals a relatively cloud-free hot spot, partially covered by Europa's cold shadow just above Jupiter's equator near 7 degrees north latitude and 325 degrees west longitude. Neon green paints cooler, tropospheric clouds. Blue colors the icy hazes that shroud Jove's upper troposphere and lower stratosphere.

Galileo took this image during its 7th orbit at a distance of 817,000 kilometers (510,000 miles) through the Near-Infrared Mapping Spectrometer (NIMS). The device can record up to 408 different wavelengths of electromagnetic radiation—from 0.7 microns to 5.2 microns—reflected at the same time from the same area. This range made it possible to measure both the sunlight reflected by Jupiter's clouds and the thermal radiation emitted from its hot spots. Because the heat, originating some 100 kilometers (60 miles) below the cloud decks, shows up at a wavelength of 4.8 microns, scientists refer to 5-micron hot spots. They are not all that hot by our standards—the temperature at their visible depth is about 0 degrees Celsius (32 degrees Fahrenheit)—but the normal Jovian forecast is around -130 degrees Celsius (-200 degrees Fahrenheit).

The remaining wavelengths measured by NIMS helped to reveal the composition of the clouds: because any given substance absorbs only certain wavelengths of light, whatever bounces back serves as a spectral signature—a telltale collection of gaps or bands in an otherwise continuous spectrum. Within Jupiter's cloud decks, NIMS observed spectral signatures for water and ammonia, as expected. Near the hot spots, and especially within them, though, the atmosphere was extremely dry. The relative humidities were between 0.02 and 10 percent.

The tiny NIMS instrument provided a wealth of information during the Galileo mission. In addition to studying Jupiter's atmosphere and cloud layers during the first, second, fourth and seventh orbits, it also mapped the mineral distributions and temperatures on Jupiter's moons. Before arriving at Jupiter, NIMS made observations of Venus, Earth, Luna, the asteroids Gaspra and Ida and the impact of Comet Shoemaker-Levy 9 with Jupiter in early July, 1994. The microwave-size machine weighed a mere 18 kilograms, or just under 40 pounds, and used on average about 12 watts of power—or one-tenth of what the average hair dryer consumes.

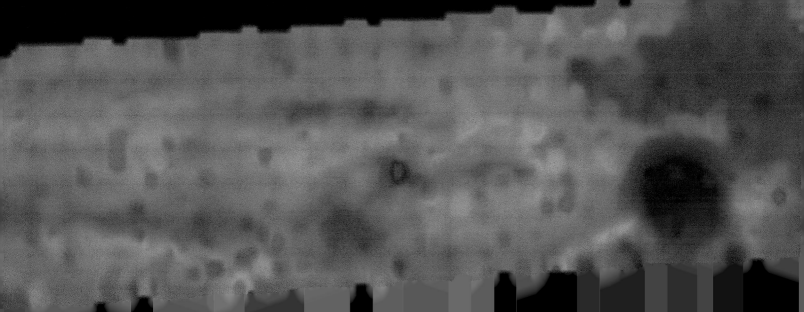

Missing Link

In yet another of Aesop's fables, Jupiter called together the beasts of the forest and offered a prize for which everyone brought forth the most handsome offspring. When a monkey presented her hairless, wrinkly son at the baby parade, Jupiter and the other animals guffawed. But the mother monkey had the last laugh. In Ambrose Bierce's account in *Fantastic Fables*, she proclaims "It is all very well to laugh at my offspring, but you go into any gallery of antique sculpture and look at the statues and busts of the fellows that you begot yourself." Jupiter, the creator of the first man, quickly awarded her first prize, whispering "Sh! don't expose me!"

Ganymede, shown here in front of Jove, is in some ways the missing link in the evolution of Jupiter's four large moons. The traits of these satellites are largely dependent on their distance from their parent planet. Io and Europa are close enough to powerfully feel Jupiter's gravitational pull, which has deformed their outer layers. Callisto, in contrast, is so far from Jove that its surface probably has not changed since the birth of our solar system some four billion years ago. Ganymede, positioned between Europa and Callisto, shares traits from both neighbors: it has an odd two-toned face, roughly half light and half dark, like an ape's bare skin against black fur. The dark patches most closely resemble Callisto's ancient face, whereas the lighter regions are not unlike Europa's younger icy crust.

This false-color photograph, a composite of three black-and-white shots taken by Voyager 1 on January 24, 1979, gives a good sense of Jupiter's might. Ganymede, which is roughly the size of Mercury, looks downright puny next to some of Jove's smaller white oval storms, seen crossing the planet's southern hemisphere. Jupiter's Great Red Spot, just below the center of the frame, readily dwarfs all of other features in the orange-cast scene. Voyager 1 snapped the image at a range of 40 million kilometers (25 million miles) as it raced towards the planet at a speed of about 1 million kilometers (621,371 miles) per day. At this distance, scientists were able to monitor the Great Red Spot and confirm ground-based observations that it circled counterclockwise during a period of about six days.

Inner Moons and Rings

INNER MOONS AND RINGS Compared to the Galilean satellites, Jupiter's four inner moons are tiny, ranging in size from about 20 to 189 kilometers (12 to 117 miles) in diameter. Only one, Amalthea, was known before the Voyager missions in 1979. The Galileo spacecraft snapped three dozen images of the little moons and found that they are lumpy and misshapen. Not only do they lack the mass needed to pull themselves together into neat, spherical balls, they are probably not even completely solid. The innermost moon, Metis, shows up as a mere point of light in the upper right corner of this near-infrared mosaic, taken by the Hubble Space Telescope on September 17, 1997.